GREEN

by Melanie Mitchell

first step nonfiction

Lerner Publications · Minneapolis

MW01199113

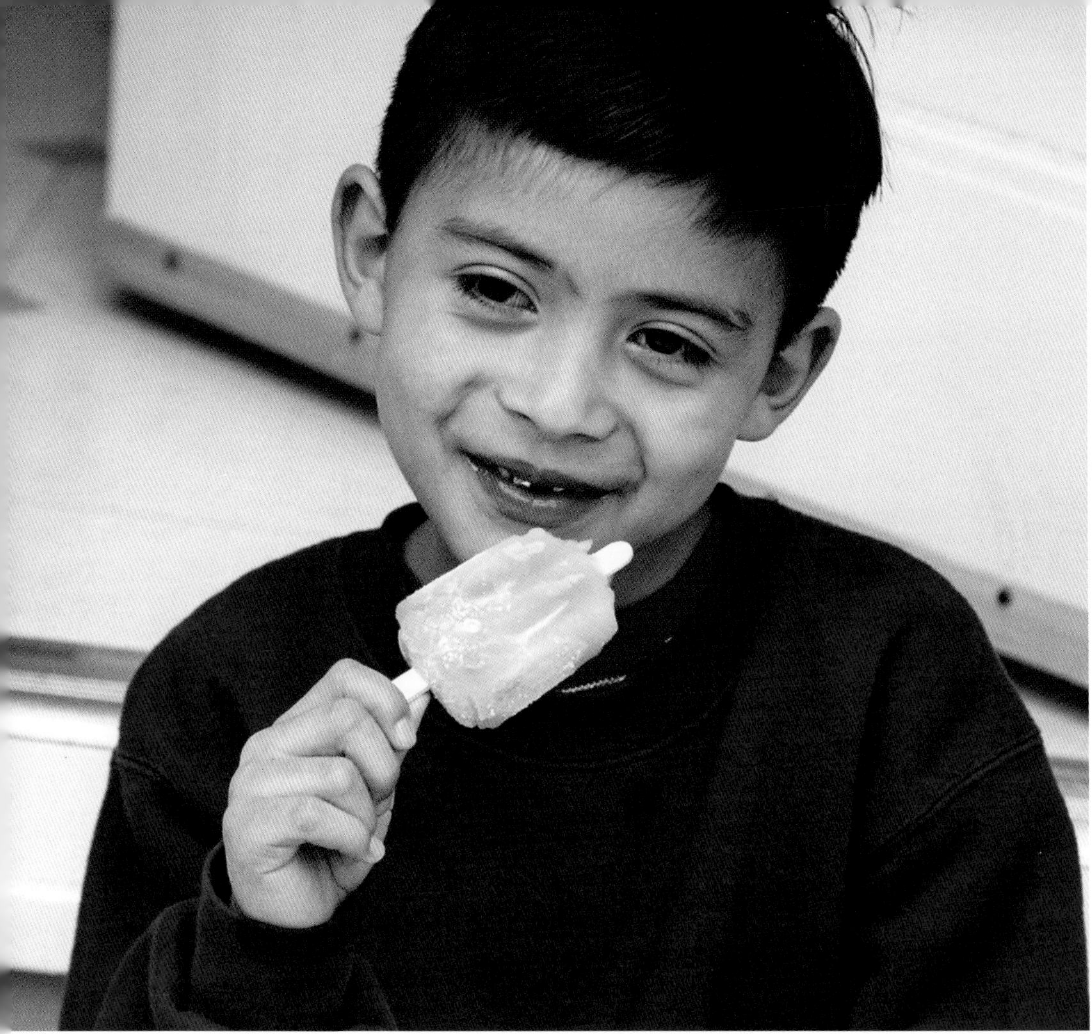

I see green.

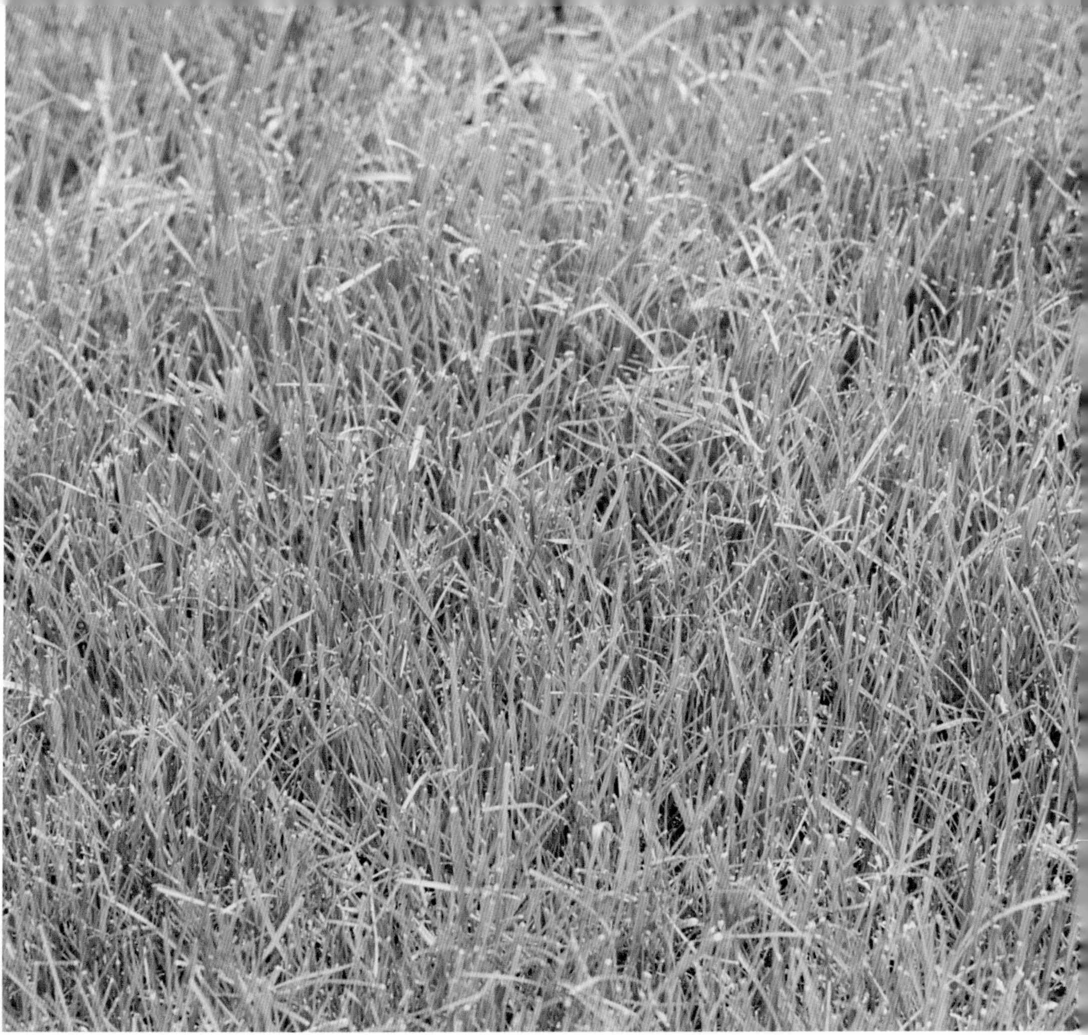

The grass is green.

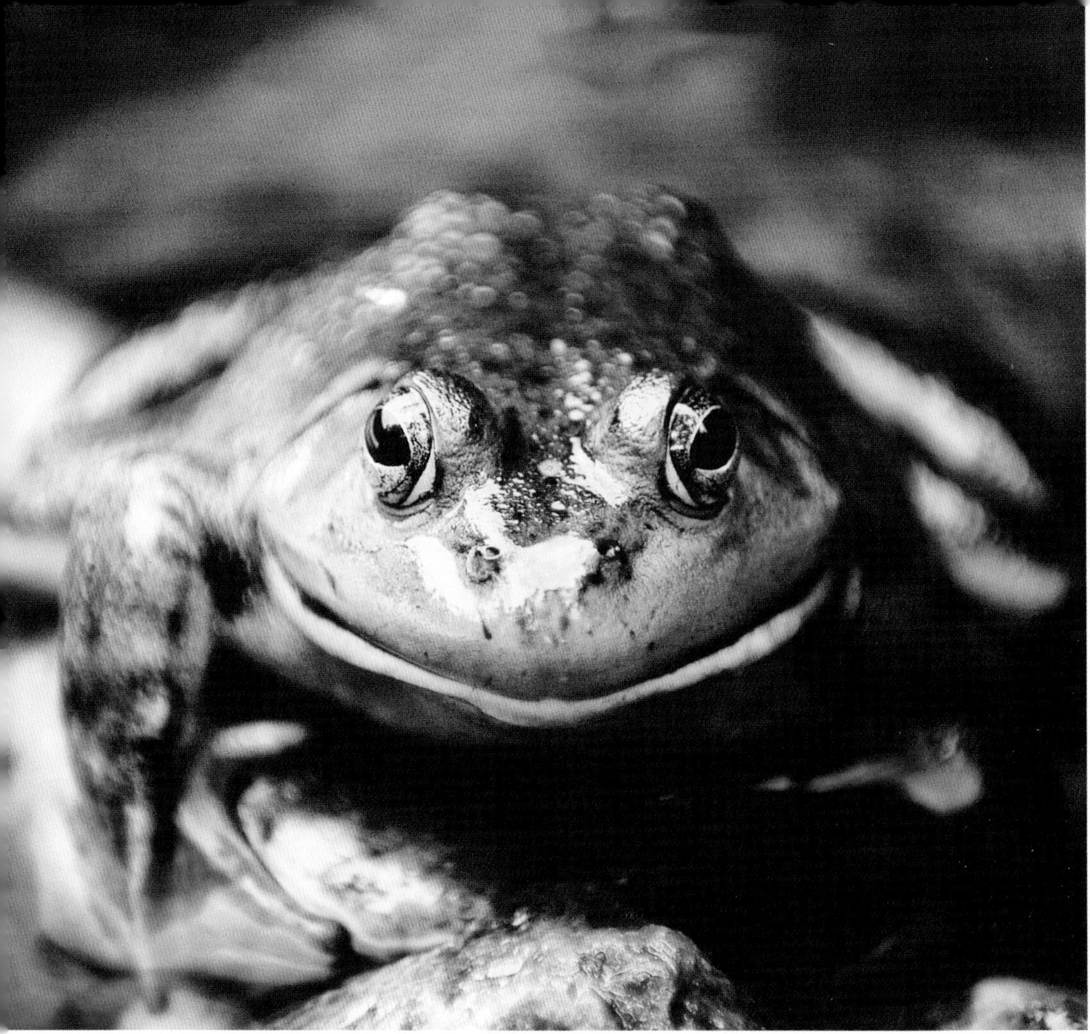

A frog is green.

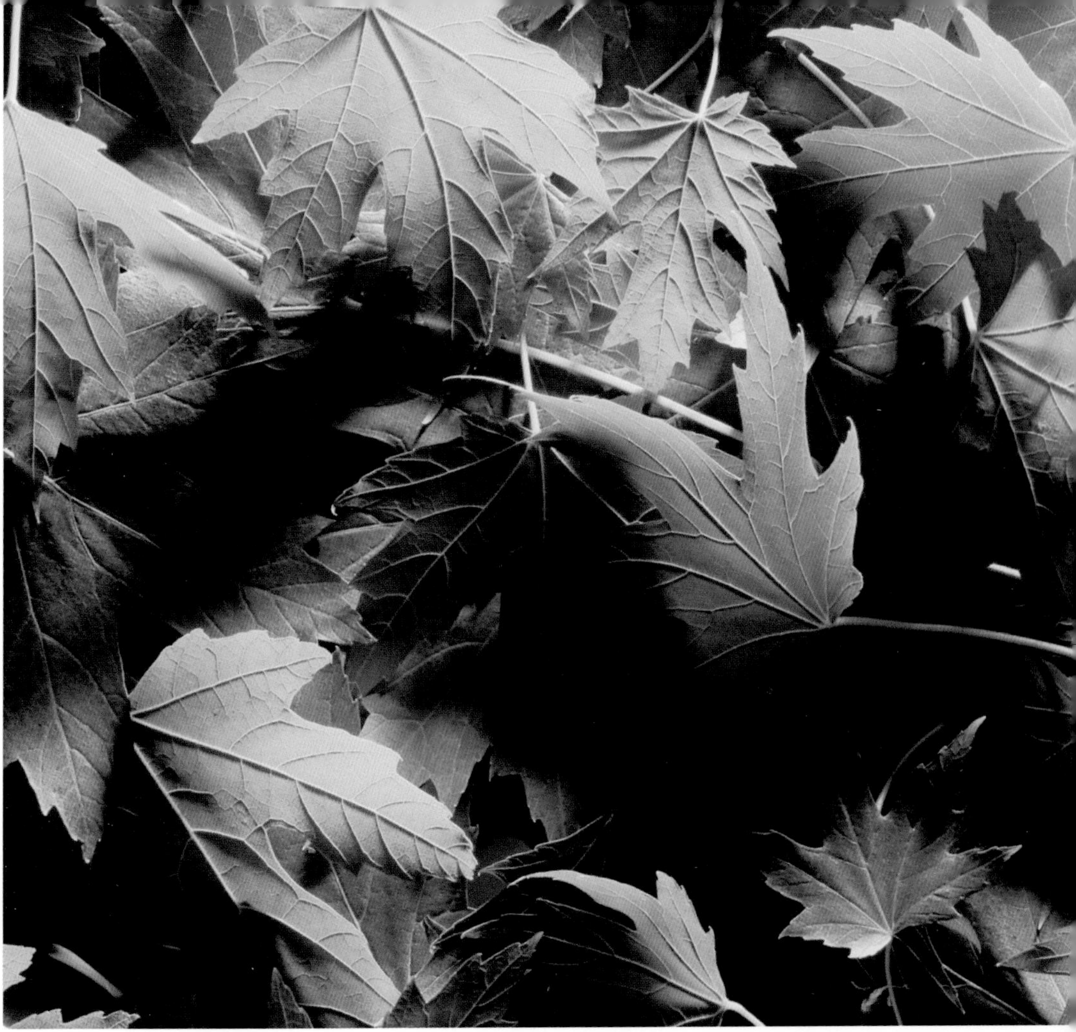

A leaf is green.

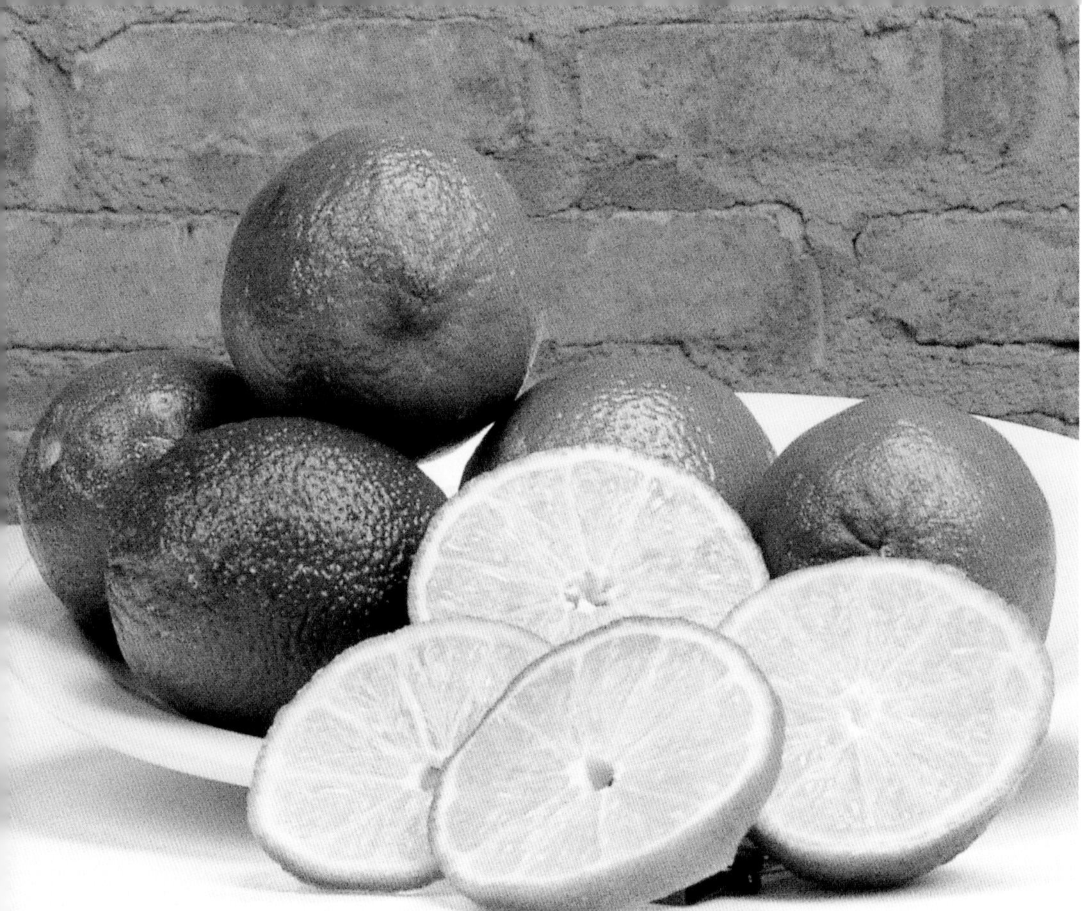

A lime is green.

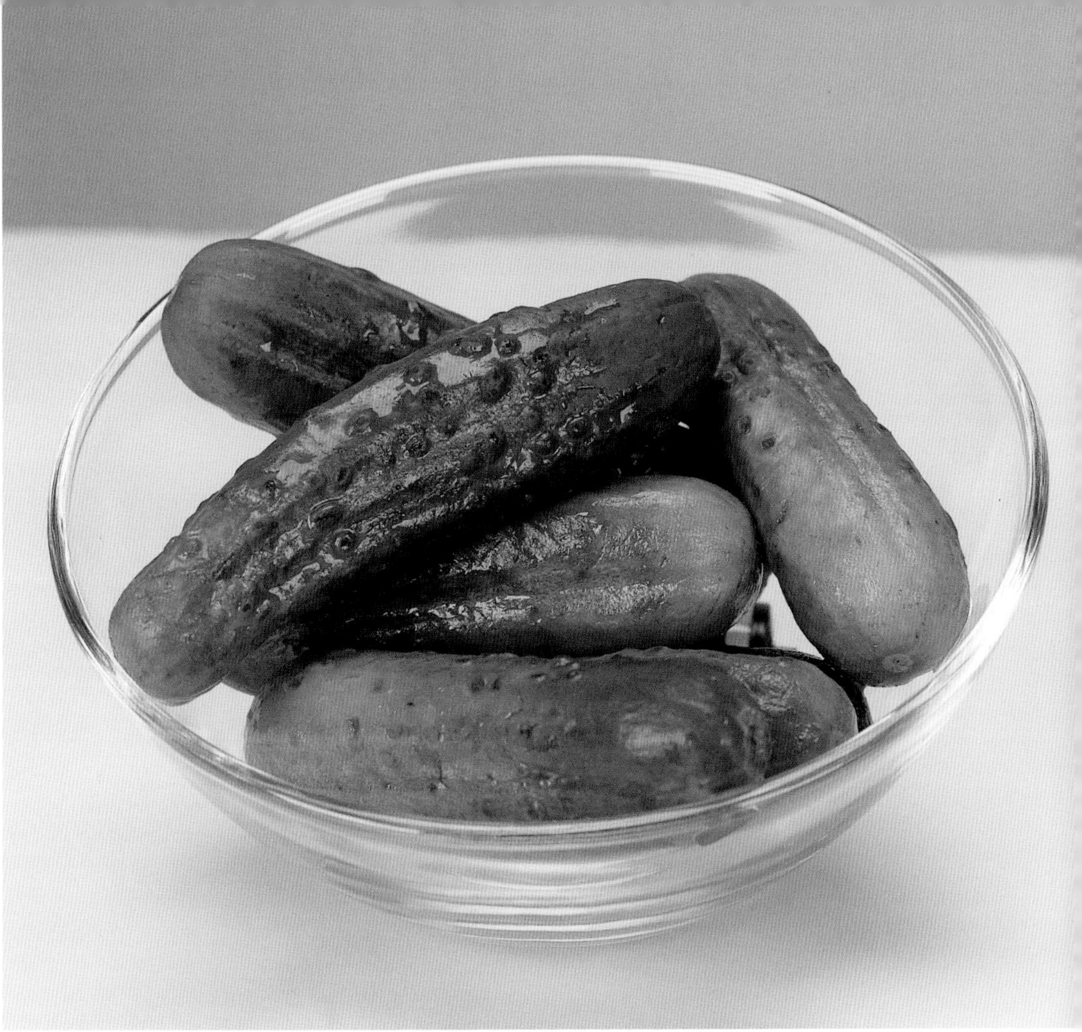

A pickle is green.

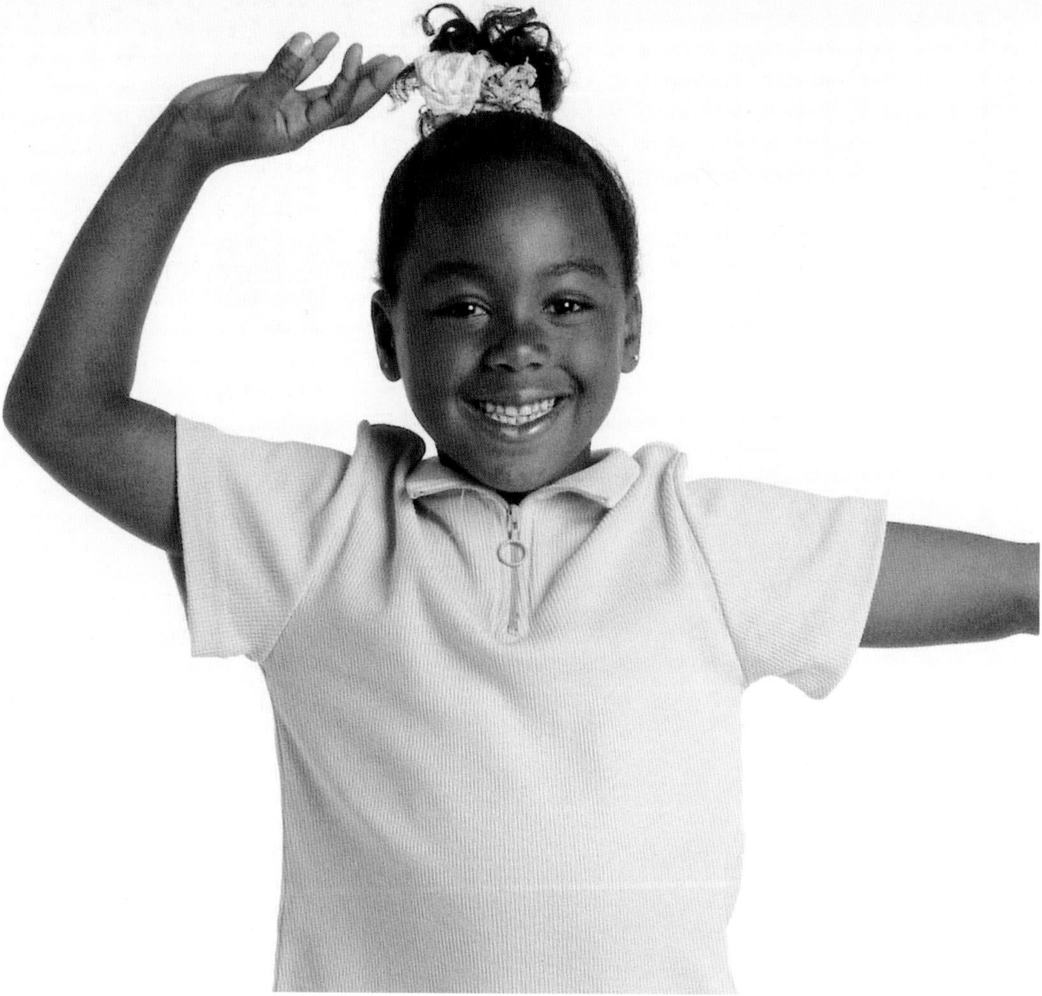

I like green!